DEDICATION

To my incredible parents,
William and Lyn Dotterweich,
and my altruistic egg donor, Diane Lubinski

This is Robin and her
husband Robert.

Robin and Robert have
always dreamed
of having babies.

When the time came to have children, Robin and Robert were so excited.

They prepared the nest
and Robin laid her eggs.

Sadly, Robin and Robert watched as all their friends had babies, but Robin's eggs never hatched.

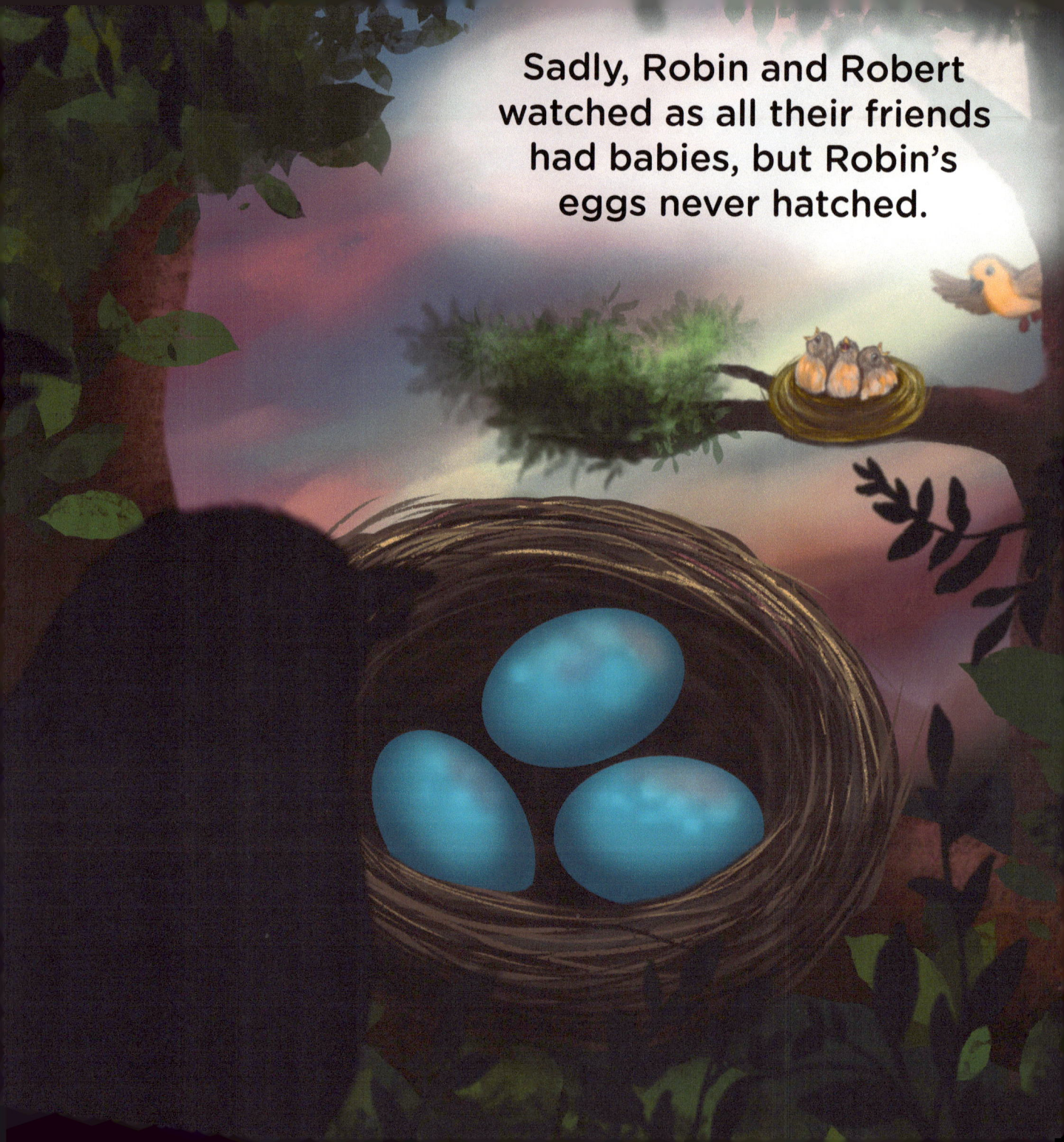

A year passed and they watched as all their friends' eggs hatched into baby birds, while their eggs never hatched.

One day a very special bird came to Robert and Robin and offered her own eggs to them.

"I have many eggs but I'm not ready to start a family yet," she explained.

"So I want to give my eggs to a couple who cannot make a family with their own eggs."

This special birdie who gave her eggs to Robert and Robin was called their "egg donor."

Robert and Robin were
delighted and accepted
the gift!

Robin sat on her nest for thirteen whole days and nights.

On the thirteenth day their babies hatched! Robert and Robin were so excited to finally be parents!

They cared for their new babies by feeding them and keeping them warm.

As the babies grew bigger and stronger, Robert and Robin taught them how to fly and how to find food on their own.

Robert and Robin were very grateful
to that special birdie who donated
her eggs to them! They had been
given the gift of parenthood.

They loved their babies
very much and their
babies loved them.

It did not matter that Robert and Robin's babies did not hatch from Robin's eggs, but instead it mattered that Robert and Robin had given them a safe home, cared for them, and taught them how to be polite, strong, and independent birds.

ABOUT THE AUTHOR

My name is **Kathryn Dotterweich** and I am graduating from the University of Virginia, Class of 2021, with a Bachelor of Arts in both neuroscience and biology. I will be attending medical school in the fall of 2022 where I hope to pursue reproductive endocrinology, maternal fetal medicine, or neonatology. I am an avid equestrian and compete at many AA rated horse shows up and down the East Coast. I also enjoy spending time with my friends, volunteering at the OBGYN unit at the Martha Jefferson Hospital in Charlottesville, VA, and fostering dogs and puppies through Punta Santiago Dogs and the local SPCA in Charlottesville.

I am an only child born in 1998 to my parents William and Lyn Dotterweich with the help of an anonymous egg donor. At this time, anonymous donation was really the only option. My parents did not tell me that I was conceived via egg donation until I was 20 years old, which came as a huge shock. After finding out I was donor conceived, I did AncestryDNA and through a second cousin and lots of internet snooping, I was able to locate my egg donor, Diane Lubinski. Diane was thrilled to hear from me, as she has always wondered if her egg donation had been able to "give another family the gift of a child."

The top reasons my parents cited when I asked why they did not tell me earlier that I was donor-conceived was that they didn't know how or when to tell me and that egg donation is not common so they were afraid of what other people might think. Therefore, I have set out on a journey to normalize egg donation and help other parents tell their children at a young age. My biggest hope is that more parents will tell their children that they are conceived via egg donation very early in development so that it does not come as a shock when they are older. Even though the child will not understand the actual science behind sexual reproduction and egg donation, knowing that they are from an "egg donor" will make it easier to comprehend when they are older. This book is a way to help teach children how they were conceived in a cute, fun, and friendly way!

ABOUT THE EDITOR:

Dania Halperin grew up in San Diego, CA, graduated Summa Cum Laude with a B.S in Biology from Landers College for Women in 2019 and currently attends the Sackler School of Medicine in Tel Aviv. Dania has worked as editor on other projects such as MCAT Biology II By MCAT King which provides quality study material for aspiring medical students. She is currently a Chairwoman of the National Premedical Association (NPreMA).

NPreMA is a 501(c)(3) nonprofit organization built by medical students to foster a supportive community of pre-medical students nationwide. Nprema aims to inspire and help create the next generation of physicians. We encourage you to join us by subscribing and donating. For more information, please visit https://www.nprema.org/.

NATIONAL
PRE-MEDICAL
ASSOCIATION

Text and illustrations copyrighted by NPREMA.org | National Pre-Medical Association | Editor: Dania Halperin | Illustrator: Cassia L. Rand | All Rights Reserved | Books can be purchased in bulk by contacting Info@nprema.org

Faculty Advisor: Dr. Noble Zaghi

ISBN: 978-1-7339906-2-2